S0-DQY-993

DATE DUE		
MAR 4	Danny	C-1
MAR 21	Andrea	B-1
SEP 29	Brenda	C-1
	Cristopher	D-2
MAR 10	Mariah	D-2
	Francis	D-2
MAR 22	Gomez	E-2
NOV 13	Lauren Sutton	P-4
JAN 11	Kathy	E-3
FEB 12	Monique	B-3

GAYLORD PRINTED IN U.S.A.

Cerra Vista School
2151 Cerra Vista Drive
Hollister, California 95023

What Is the Moon?

A **Just Ask** Book

by Chris Arvetis
and Carole Palmer

illustrated by James Buckley

Copyright © 1987 Checkerboard Press, a division of Macmillan, Inc.
All rights reserved. Printed in U.S.A.
Library of Congress Catalog Card Number: 87-70370

CHECKERBOARD PRESS and colophon and JUST ASK
are trademarks of Macmillan, Inc.

CHILDRENS PRESS CHOICE

A Checkerboard Press title selected for educational distribution
ISBN 0-516-09868-3

Look at the big bright circle in the sky.

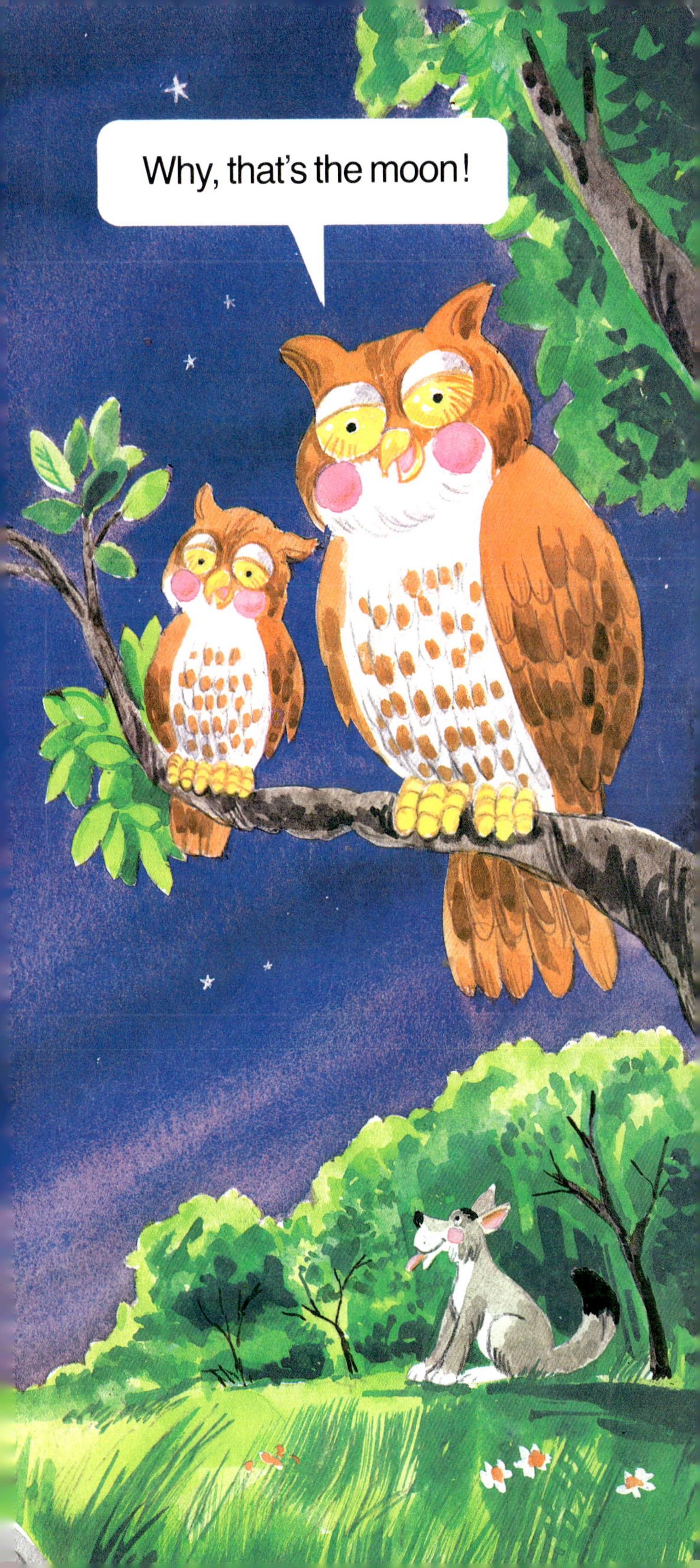
Why, that's the moon!

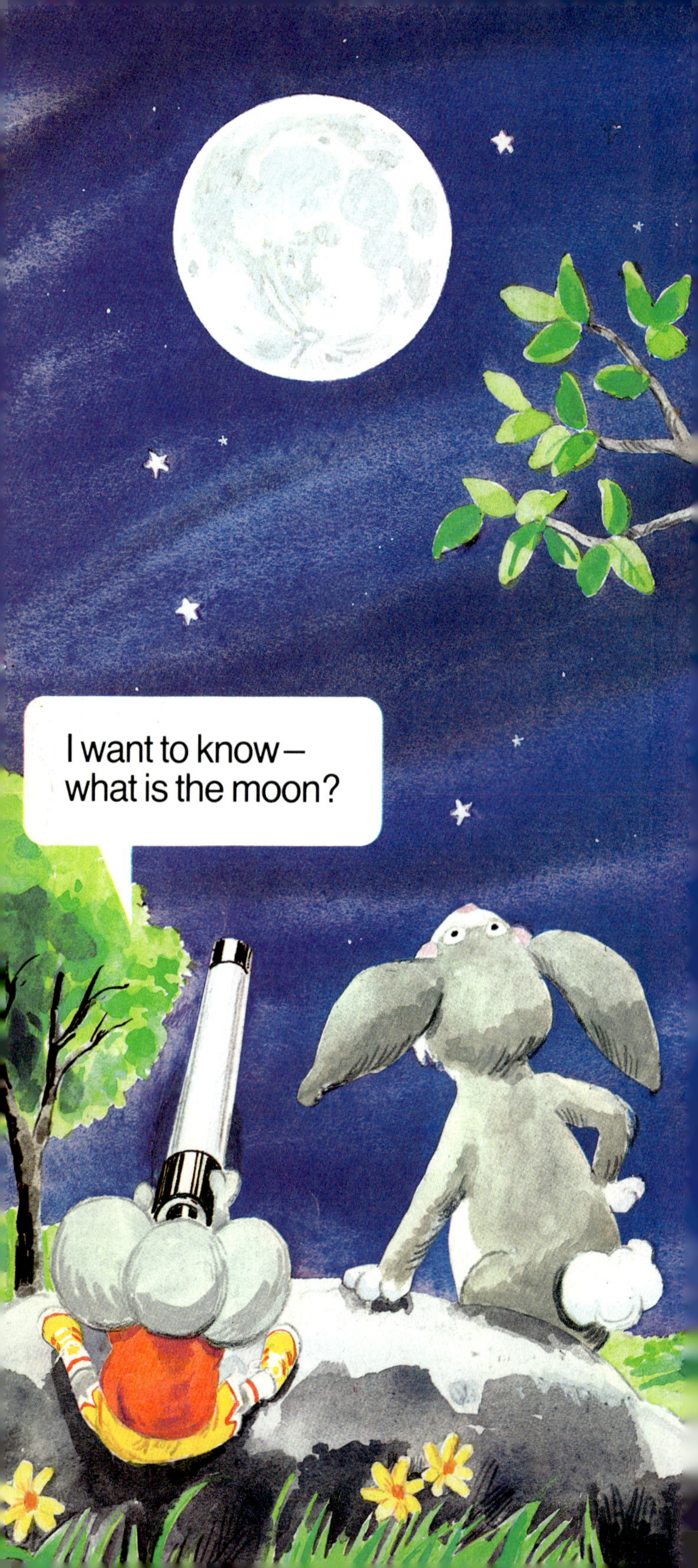
I want to know–
what is the moon?

Well, the moon is a huge ball of rock. It lights up the sky at night.

The moon looks like a round, gray ball with some darker gray patches.
These dark gray patches are flat land.

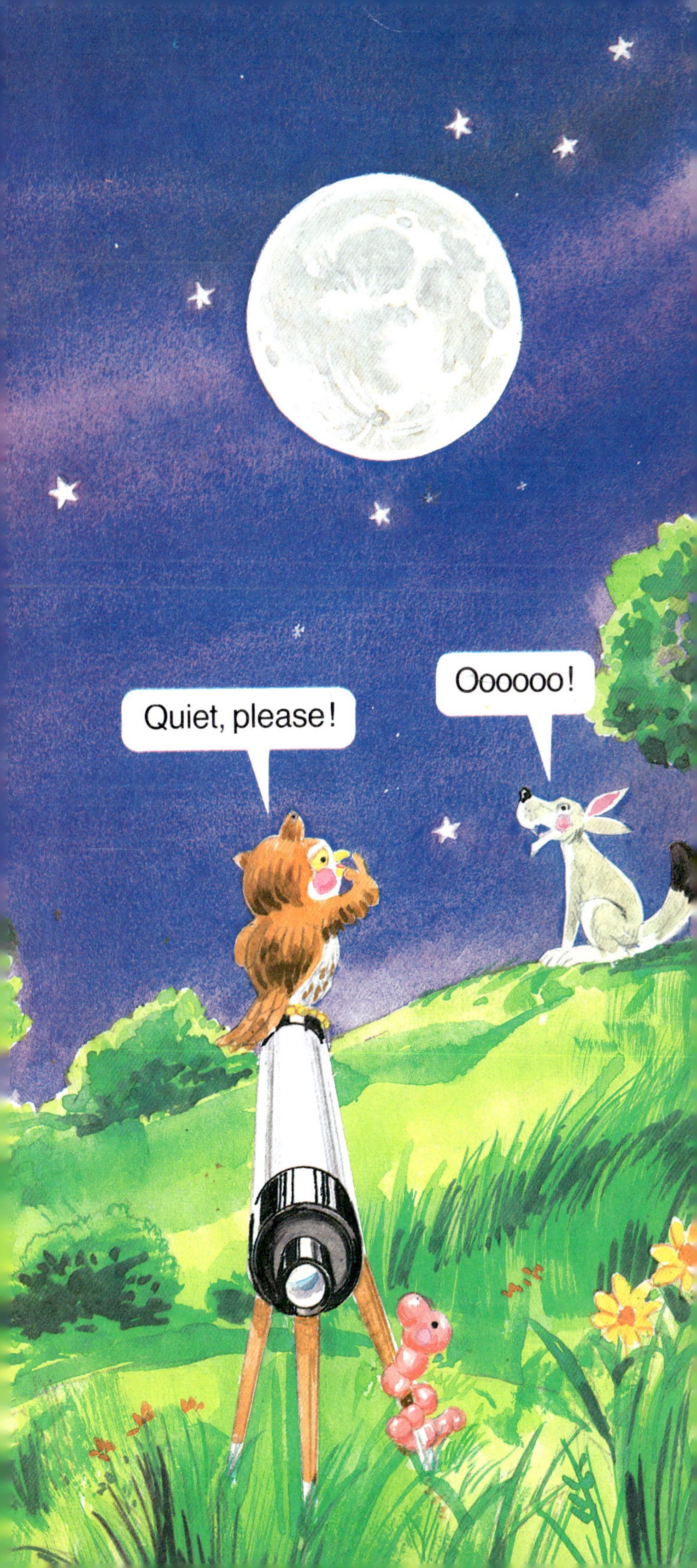
Quiet, please!
Oooooo!

"I see mountains!"

Look at my picture.
The rest of the moon is rough.
It has big mountains.
These parts are called highlands.

The large dents that look like bowls are called craters.
That's CRA-TERS!

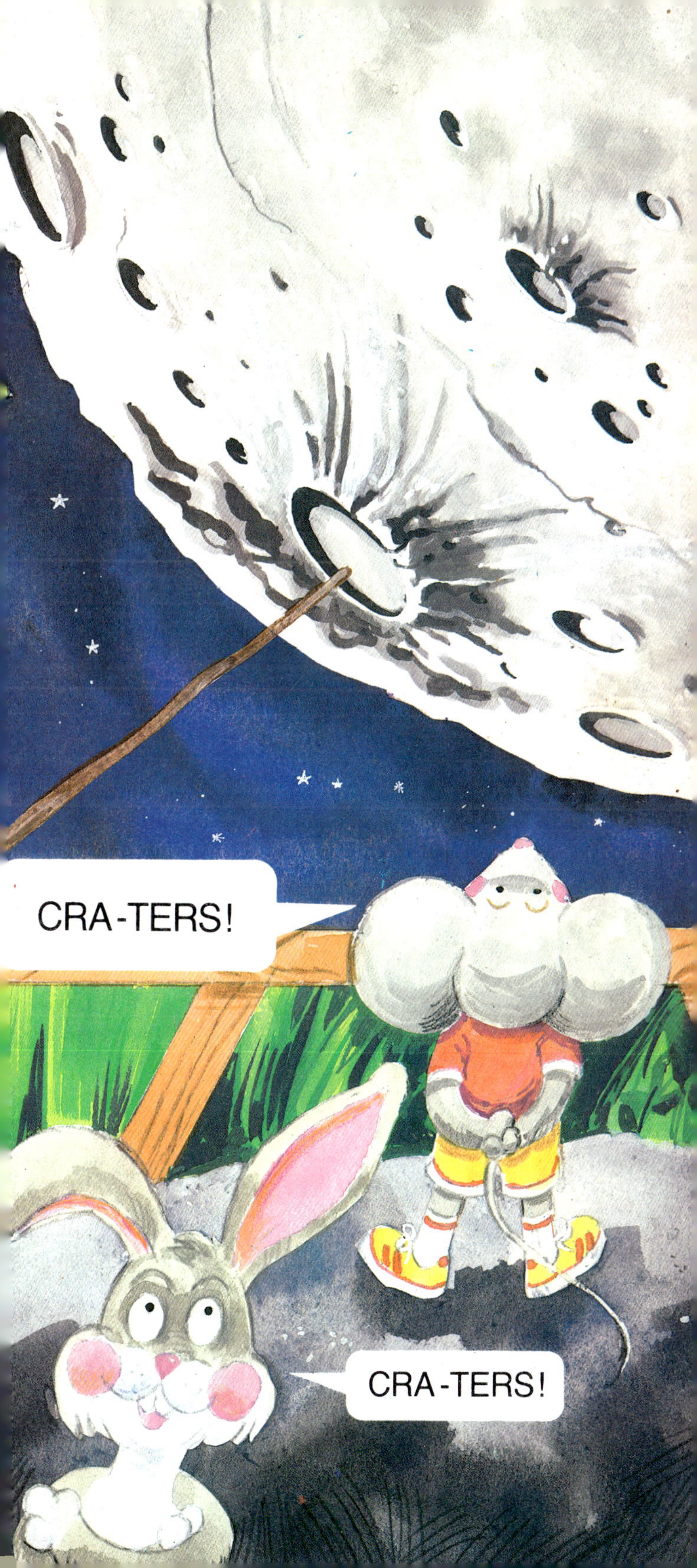
CRA-TERS!
CRA-TERS!

The craters are made when hard things hit the moon. Some of the craters look like volcanoes.

"You have big eyes!"

See those cracks?

The moon has valleys, too.
The long narrow ones look
like big cracks in the moon.

I'm thirsty!

There are also winding valleys that look like dried-up rivers.

The moon goes around the earth.
It follows a path like this.
It takes about a month
to make one trip.

One month!

For one trip?

Astronauts have traveled to the moon.
They have walked on the moon.
Since then, we know a lot more about the moon.

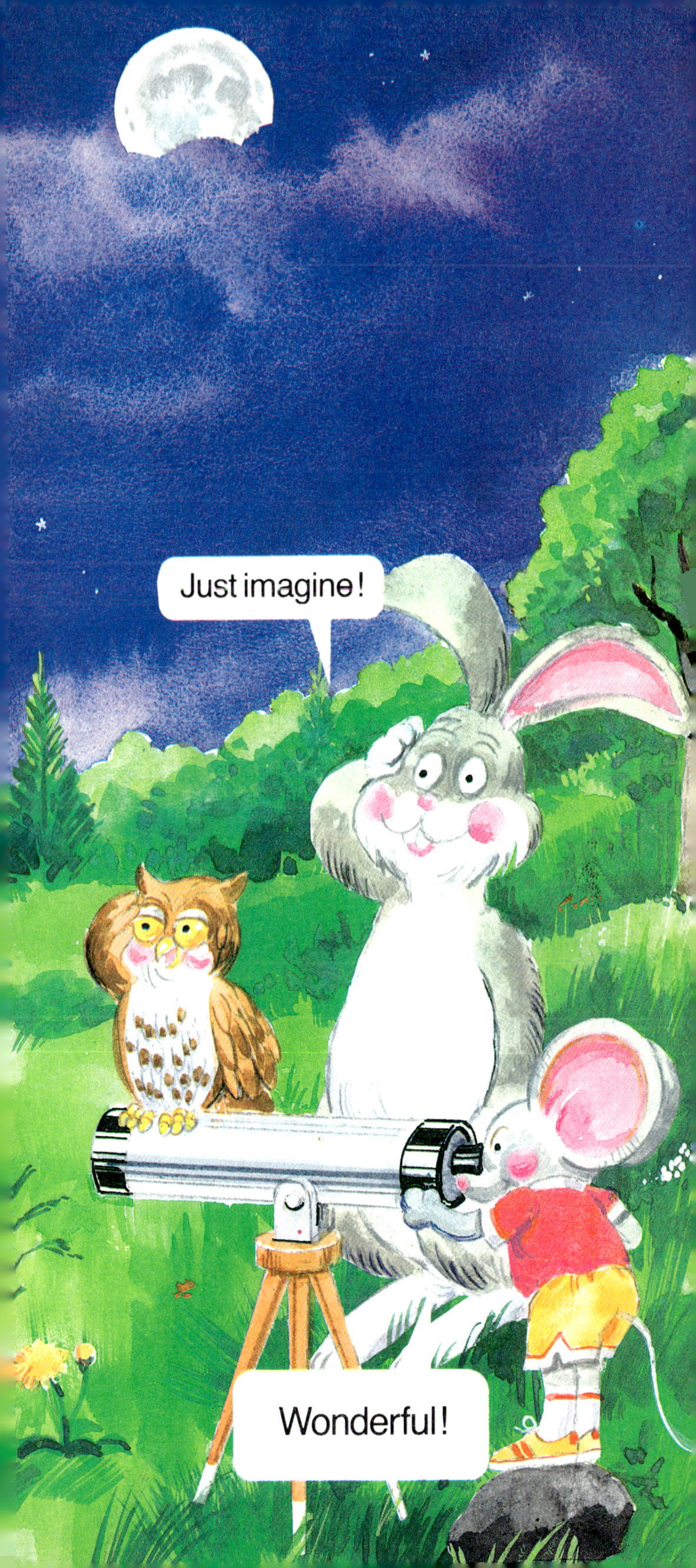
Just imagine!
Wonderful!

We know that it is very quiet and lonely on the moon.
There are no plants, animals, or people.

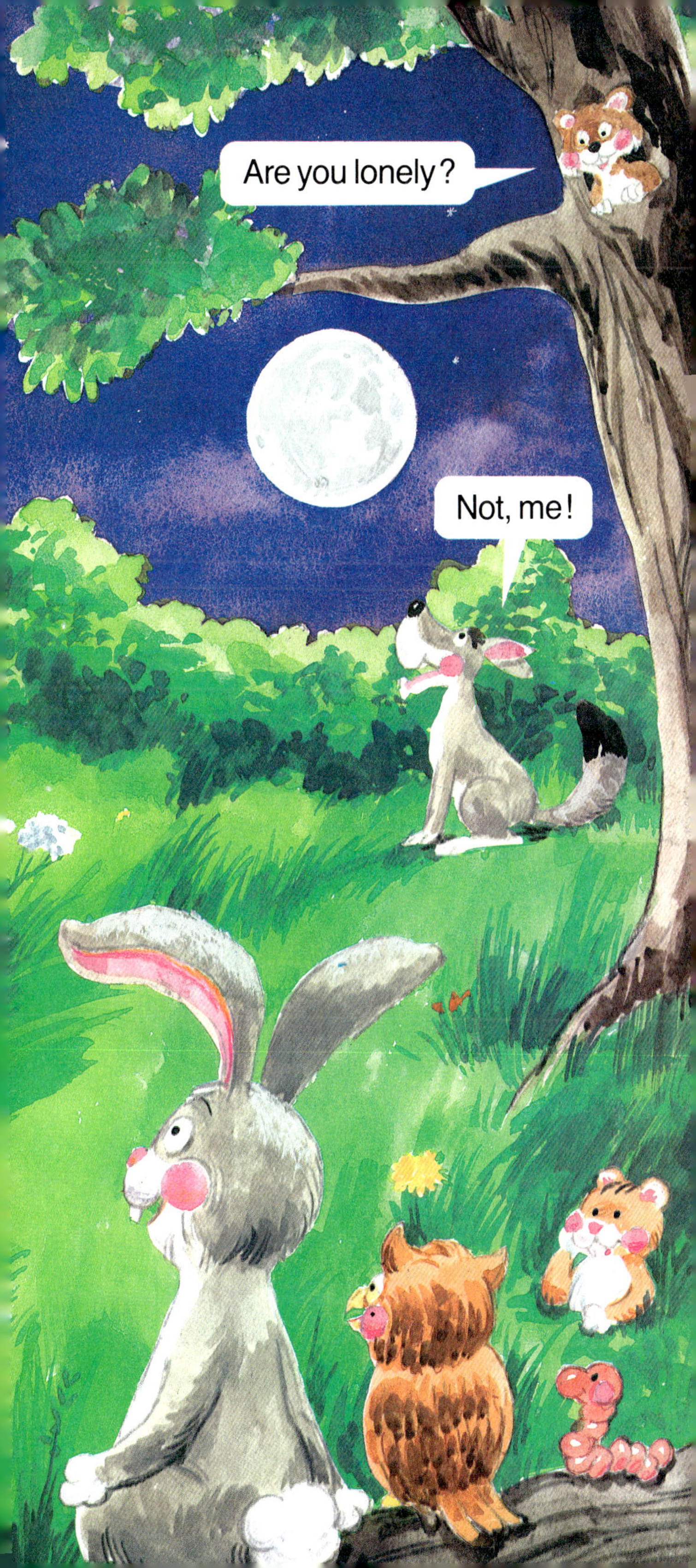
Are you lonely?
Not, me!

There is no air or wind on the moon.
You could not fly a kite on the moon.

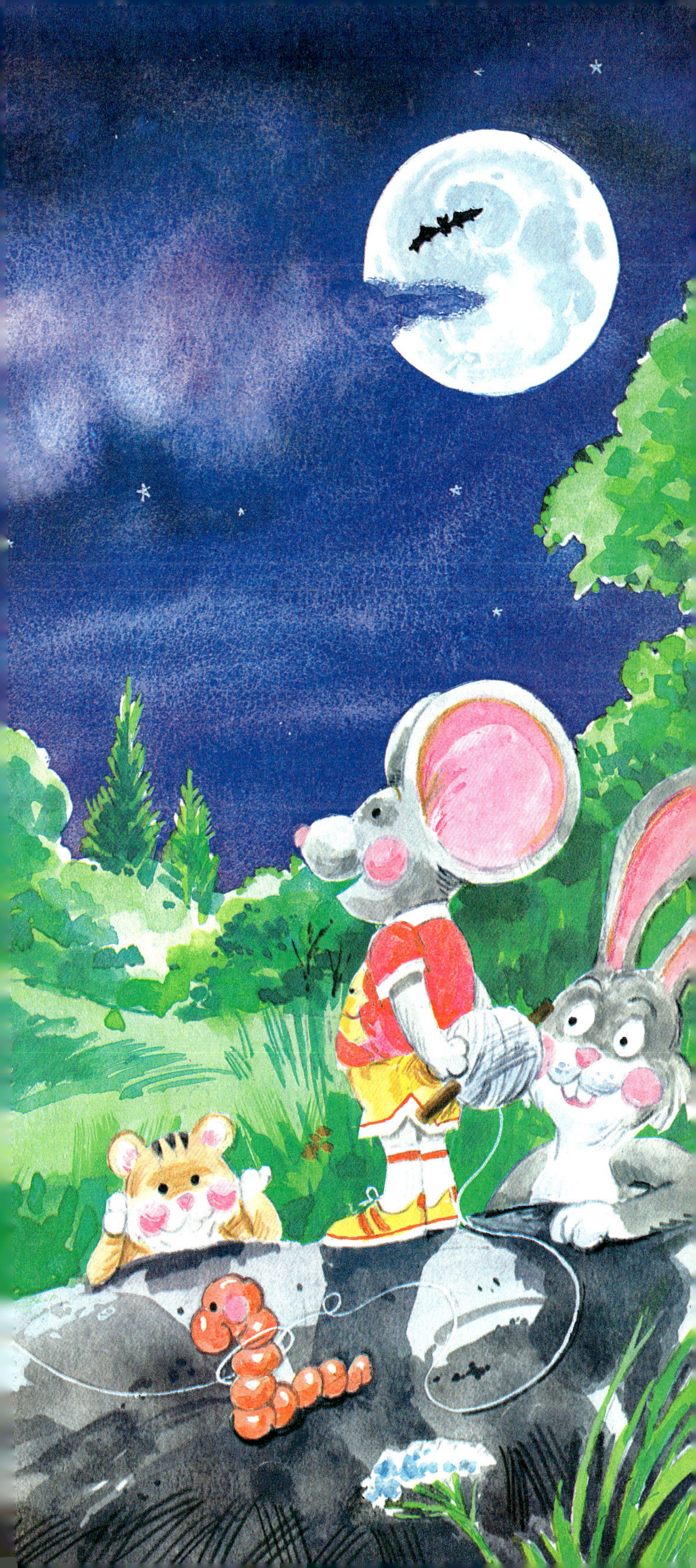

There is no water.
The temperature is a lot hotter and colder than anywhere on earth.

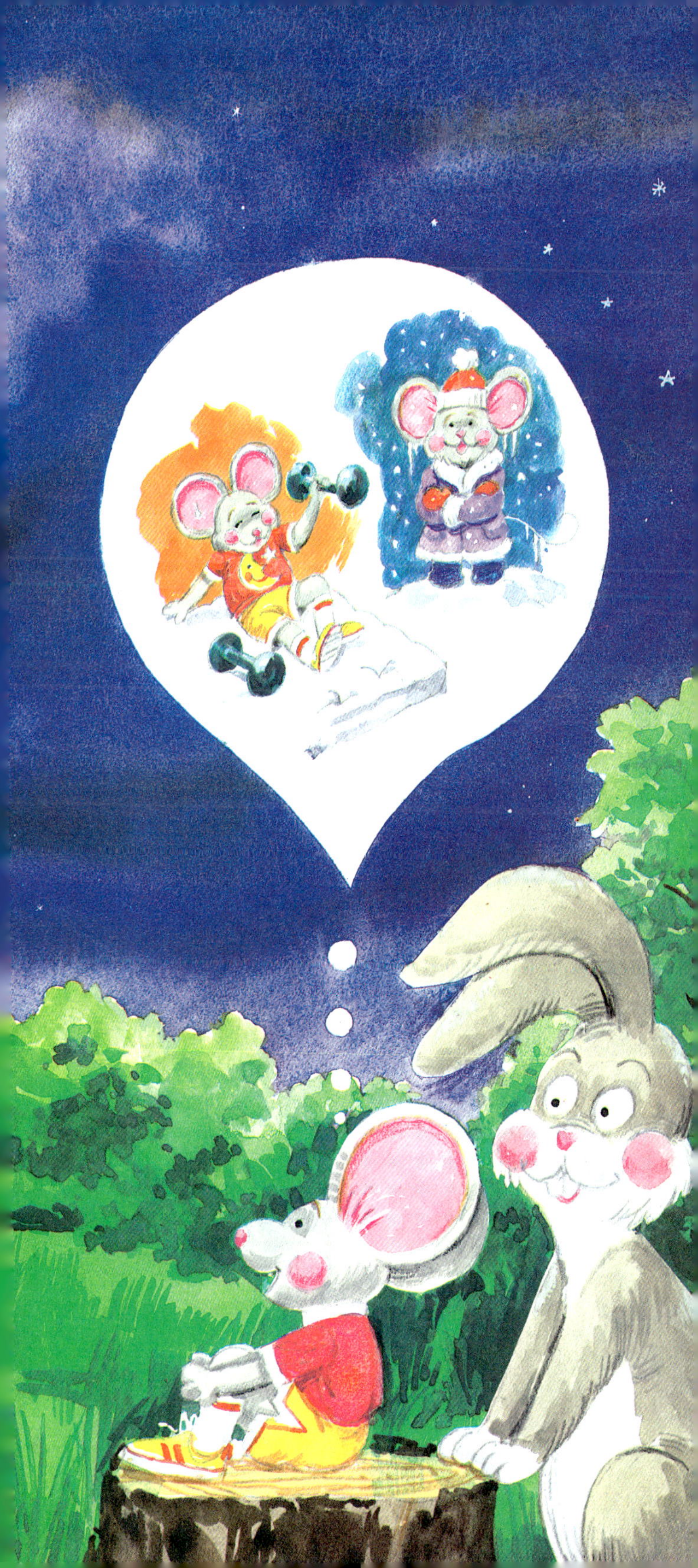

You would have to wear a spacesuit if you traveled to the moon.

You would be much lighter on the moon.

You would float in space and bounce as you walk.

Like this?

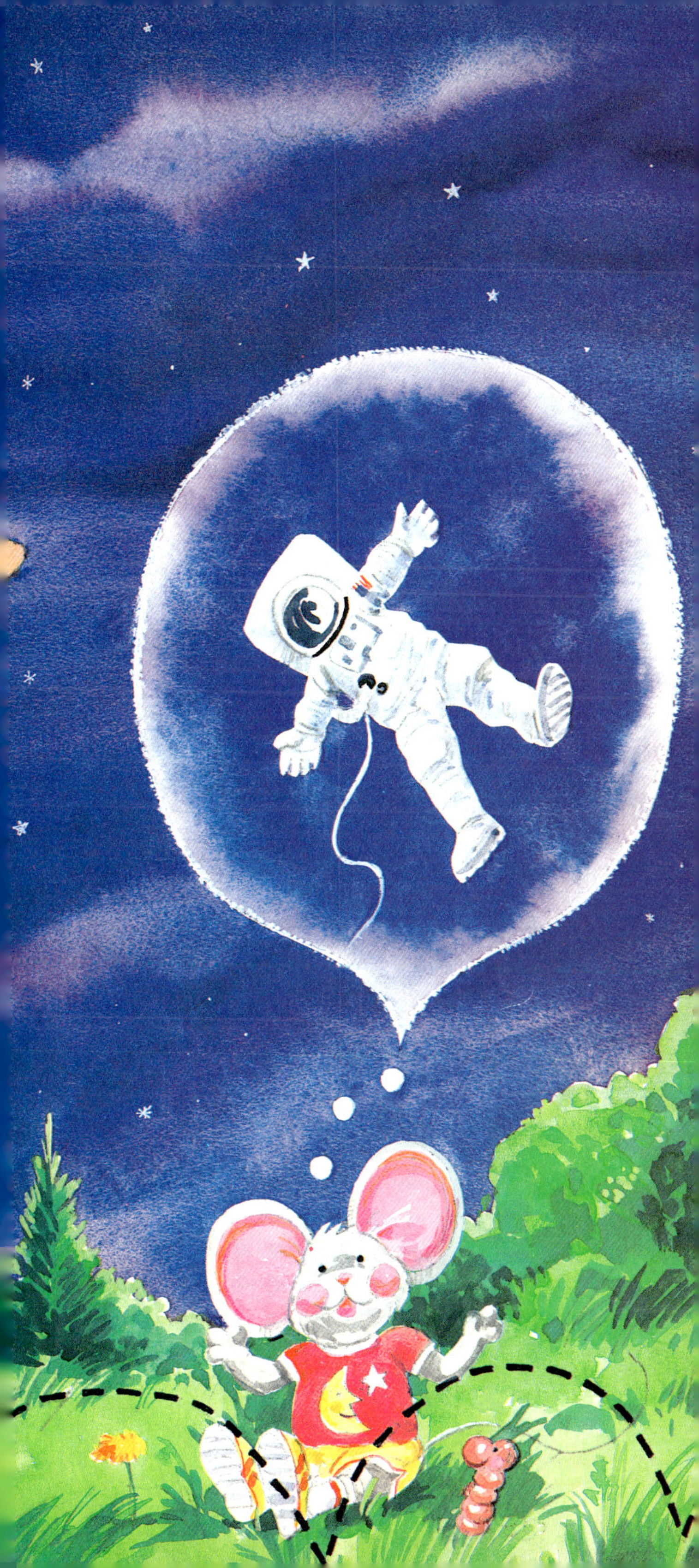

As people study about the moon, we will know more about it.

Some day, the moon may be a space station for space travelers.

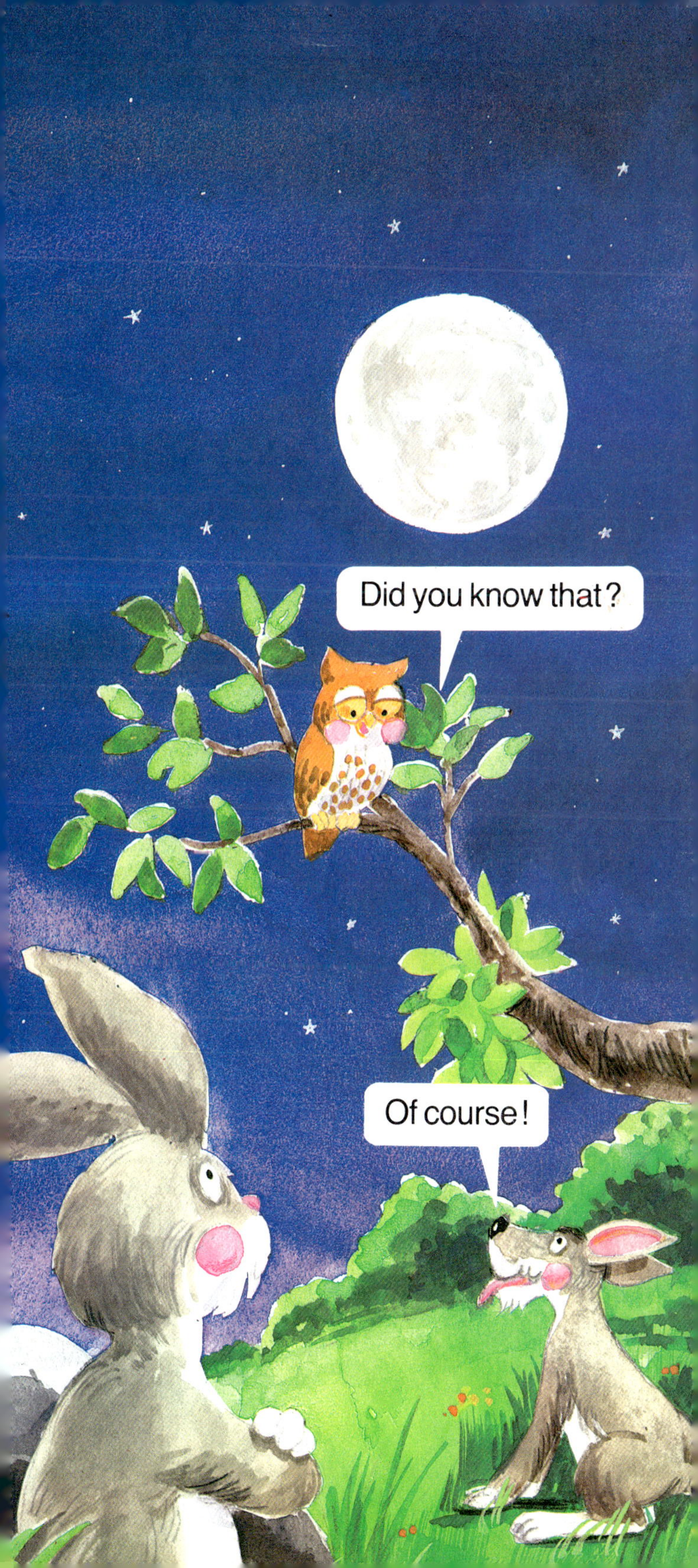
Did you know that?
Of course!

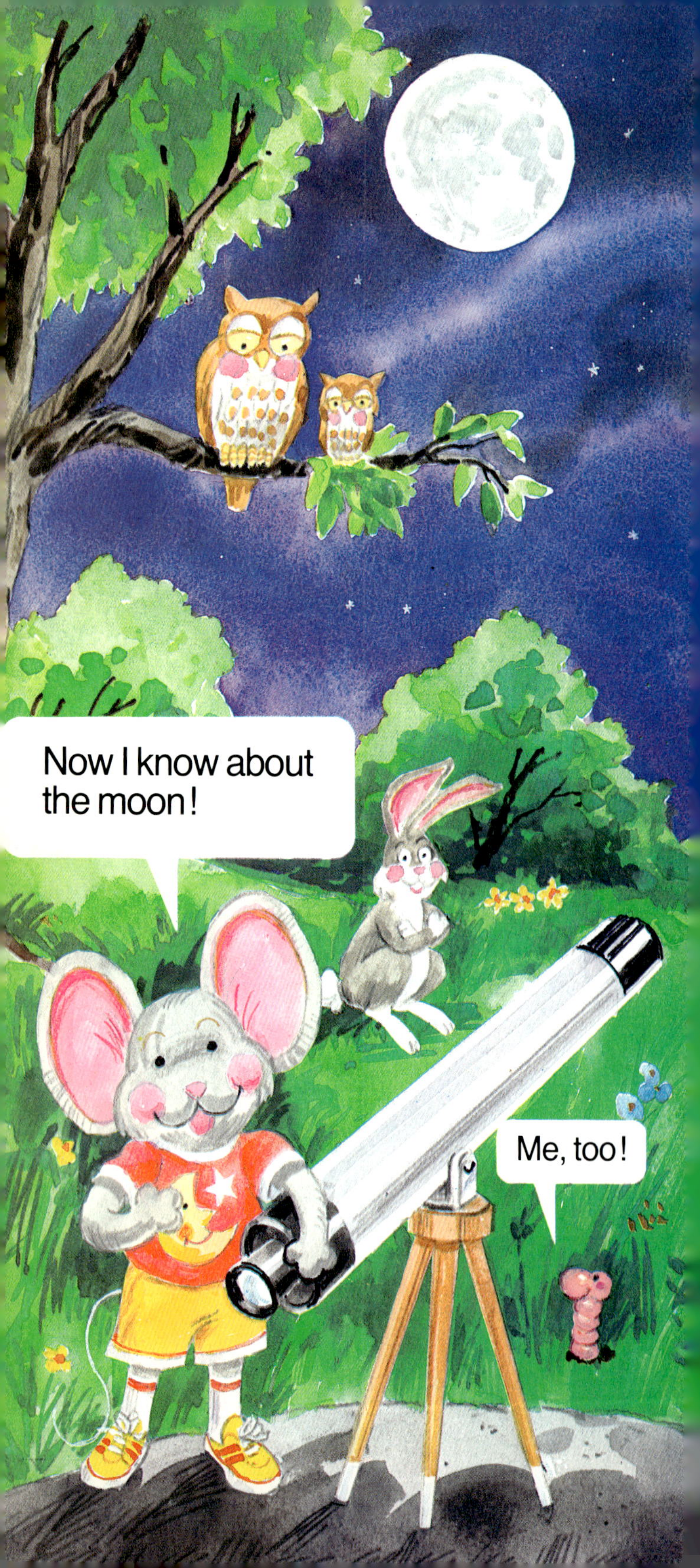
Now I know about the moon!
Me, too!

Cerra Vista School
2151 Cerra Vista Drive
Hollister, California 95023